MONARCHS OF THE MIST

THE STORY OF REDWOOD NATIONAL PARK AND THE COAST REDWOODS

BY JOSEPH E. BROWN

STAY AWHILE. HIKE SOME OF THESE PILLARED PATHS.
STAND AT THE EDGE OF A MOONLIT PRAIRIE.
CAPTURE WILD ELK ON FILM. WALK A CANYON
UPHOLSTERED IN A TRILLION FERNS.
EXPLORE THE COASTLINE'S GRAY-MISTED MILANGE
OF SOARING HEADLANDS AND DRIFT-LOG JUMBLES.
LINGER BENEATH THE WORLD'S TALLEST TREES—
CLOUDSWEEPING MONARCHS OF UNFORGETTABLE BEAUTY.
REVEL IN THE AMBIENCE, AND WHEN YOU LEAVE,
THE MAGIC WILL GO WITH YOU.
Sandra L. Keith

ISBN 0-911235-00-0

Howland Hill Road.

PREFACE

Redwood National Park was established in 1968 and expanded in 1978. Both of these congressional acts were in response to public interest in protecting some of the remaining virgin redwood lands from the direct and indirect effects of timber harvesting.

Of the 58,000 acres within the 1968 legislative boundary, nearly half were within three California Redwoods State Parks: Prairie Creek Redwoods, Del Norte Coast Redwoods, and Jedediah Smith Redwoods. These three parks are administered separately by the California Department of Parks and Recreation; national park lands are administered by the National Park Service. Nonetheless, the goals of each agency are similar: the perpetuation and restoration of some of America's finest natural features for the public inspiration, enjoyment and scientific study.

The forest, stream, and seashore features of the state and national parklands are merged into one majestic setting, and the natural features and processes described here do not differ in significant detail from park to park within the national park boundary.

The state redwood parks and their major features:

Prairie Creek Redwoods State Park. Largest of the three state parks at 12,240 acres, it is located north of Orick on Highway 101. This park possesses dense forests, consisting largely of 300-foot high old-growth redwoods; six-mile long Gold Bluffs Beach; and Fern Canyon, a 50-foot deep cleft where delicate, five-fingered ferns proliferate.

Del Norte Coast Redwoods State Park. Covering 6,400 acres bordering the Pacific Ocean, it is noted for its spectacular seacoast vistas, thick stands of old-growth redwoods, and beautiful rhododendrons. It extends south of Jedediah Smith Redwoods State Park to False Klamath Cove.

Jedediah Smith Redwoods State Park. Its 9,200 acres include the 5,000-acre National Tribute Grove of old-growth redwoods, honoring those who served their country in World War II; the Stout Memorial Grove; and the historic Howland Hill Road. The park is located along the Smith River, off Highway 199 north of Crescent City.

Redwood National Park is open all year, and its activities range from hiking and camping to beachcombing and fishing. Visitor information, as well as publications which contain more specific information on topics covered in this general book, are available at park visitor centers located in Crescent City, Hiouchi, and Orick.

Humans are dwarfed by redwoods. Vanilla leaf ground cover (opposite).

REDWOOD COUNTRY

Some call it the Redwood Belt. Others refer to it as Redwood Country or the Land of the Giants. Many residents proudly describe it as the Redwood Empire. Bathed in winter rain and shrouded in life-sustaining summer fog, this 450-mile long corridor of north coastal California is the present domain of the majestic coast redwood, *Sequoia sempervirens*.

Redwoods were once distributed over vast regions of the earth. Fossil records indicate that millions of years ago they were found in Europe and Asia as well as in North America. Some grew as far north as St. Lawrence Island on the Bering Sea, a place far too inhospitable today for redwood growth—but in times past, the site of forests perhaps as beautiful as any in today's California.

But physically and climatically, the earth changed. About 28 million years ago the Rockies, Sierras, Alps, and Andes began pushing skyward and as they rose, the North American continent succumbed to a drying trend. The loss of the subtropical atmosphere, coupled with the last great Ice Age, created cooler conditions which no longer favored the coast redwoods. Today its range extends naturally only from southwestern Oregon's Chetco River (where there are two groves), south to Salmon Creek in Monterey County, California. The narrow belt averages 30 miles in width and covers approximately 2 million acres (of which a scant 10 percent remains in old-growth redwoods).

Geologically, California's north coast is considered to be among the most scrambled in the United States, and only recently have scientists begun to fully understand and interpret it.

Rocks of the Franciscan Assemblage underlie much of coastal northern California. Formed from sea floor sediments, these rocks consist mostly of muddy sandstones, though some

"HEAVEN IS UNDER OUR FEET AS WELL AS OVER OUR HEADS."
—Henry David Thoreau

volcanic and metamorphic rocks are also present. There are few fossils.

The Franciscan Assemblage is the region's major geologic feature and originated as a result of a prolonged head-on collision between the Pacific's eastern sea floor and the edge of the North American Continent. As these two plates began colliding (about 160 million years ago), the accumulation of sediments was greatly deformed and slowly rose to form the present Coast Range. As a result of their complex history, these rocks are highly susceptible to weathering and erosion.

Within Redwood National Park's borders, the Coast Range rises from sea level to a maximum altitude of 3,097 feet. Streams ribbon this range, generally moving from the southeast to the northwest before emptying into the Pacific Ocean. In some places, the topography is marked by gently-rounded summits; in others, uplift and erosion of less-resistant rocks has created steep-sided drainages.

Thirty-five miles of varied coastline forms the park's western boundary. Here are marine terraces, sand or rock beaches, and an abundance of sea stacks—rock islets that raise jagged fingers just past the surf. And dotting the park's southern section are the lagoons (such as Freshwater Lagoon), placid pools created thousands of years ago when the sea level rose and drowned stream valleys. With time, consistent wave action built large sand bars across the small outlet streams and created the lagoons seen today.

Some of the park's beach cliffs are composed of river sediments deposited as an ancestral delta of the Klamath River two million years ago. In other places, the older Franciscan bedrock forms a rocky coastline. Sand is deposited on the beaches from inland sources by rivers as well as through wave erosion of the shoreline.

For coast redwoods, which grow best in a temperate, moist environment, the climate of California's north coast is ideal—mild, wet winters and moderately-dry summers. The park rarely experiences extreme heat or cold. Near the coast, the difference between winter and summer temperature averages only about 10 degrees. However, as one travels inland, there is a more pronounced temperature variation.

Perhaps even more important to redwoods than mild weather is coastal fog. The marine influence frequently produces a low cloud layer, or fog, which moderates the hottest, driest summer conditions. Some observers have called this "nature's air conditioning."

Like all trees, redwoods lose hundreds of gallons of water through their leaves on hot, dry days. It is a process called transpiration, and under drought conditions, can lead to wilting and death of the trees' upper foliage. Fog helps reduce this water loss, for it condenses on the foliage and drips to the ground.

The redwoods are not the only beneficiaries of the region's benevolent climate. Altogether, more than 1,000 species of plants and animals have been recorded within the park's boundaries. They include several rare and endangered species: California gray whales, bald eagles, brown pelicans, peregrine falcons, and the Aleutian form of the Canada goose.

Black bears, mountain lions, bobcats, river otters, elk, beavers, and a host of other animals—large and small—attest to the importance of the park beyond the protection of its best-noted resident, the lofty, mighty coast redwood.

Lady Bird Johnson Grove, typified by old-growth redwoods such as these monarchs, was the site of the dedication of Redwood National Park in 1968. An easily-walked trail leads visitors through the grove, located a short distance off Bald Hills Road.

Rhododendron Trail in Prarie Creek Redwoods State Park (above) affords a contemplative experience in an old growth forest; in spring the pink of rhododendrons adds vivid color to the forest scene. Opposite: thimbleberries and salmon berries grow in profusion in the quiet redwood groves.

Corkscrew Tree, Prairie Creek Redwoods State Park. Opposite: vine maple.

THE REDWOOD FOREST

They are monarchs of the mist, these loftiest of all living things on earth, reaching for the sunlight above the darkened, cathedral-like forest which they dominate. They are the coast redwoods, "kings of their race," as naturalist John Muir called them—for they are living links to the age of dinosaurs, and even today are among the most remarkable trees on earth.

Towering high above a complex domain, the coast redwood is not only a tree of regal form and beauty, but an amazingly strong and hardy species as well. Once it has survived the perils of seedling and sapling years and acquired its thick armor of burnt umber (which weathers to gray), it seems virtually indestructible. Here is a tree that resists pests and animal attack. Here is a tree less vulnerable to fire than most other species. Here is a tree that grows from sprouts as well as seeds—an unusual feature for coniferous trees—and which, even after death, does not cease its contribution to its environment. To no other tree, perhaps, does the term "everlasting" so rightfully belong.

The first known European to record the redwood in its present California coastal habitat was Fray Juan Crespi, a member of the 1769 Don Gaspar de Portolá expedition which was traveling northward from the Pájaro River—now the boundary between Santa Cruz and Monterey Counties. Not recognizing the tree from his previous travels, Fr. Crespi observed its distinguishing color and noted in his diary that he had named it *palo colorado* (red tree).

Archibald Menzies, a surgeon and botanist with the later George Vancouver expedition, is credited with the coast redwood's actual botanical discovery. Menzies recorded his findings in 1794, but another 53 years passed before Stephen Endlicher, a

Hungarian botanist, named the genus and species: *Sequoia sempervirens*. The generic name is believed to honor Sequoyah, a Cherokee Indian who is credited with developing an 83-character syllabic alphabet for his people. *Sempervirens* is a Latin word meaning evergreen.

The tree that attracted the interest of Crespi, Menzies and Endlicher is a member of one of three closely-related genera of redwoods, each having a single species. The giant sequoia or Sierra redwood, *Sequoiadendron giganteum*, also occurs naturally only in California, but is both larger in volume and longer-lived than its coastal relative. The third species is the small dawn redwood, *Metasequoia glyptostroboides*, which was believed to be millions of years extinct until living specimens were discovered in a remote section of China in 1944. Although all three are conifers (trees which reproduce by seed-bearing cones), the coast and Sierra redwoods are evergreens. The dawn is deciduous and sheds its needles once a year.

Both coast and Sierra redwoods have a reddish fibrous bark, an absence of resin cells, an abundance of tannin, and a red heartwood. But while the giant sequoia exceeds in total wood volume and attains a greater diameter, the coast redwood grows taller. In fact, the world's tallest known tree, a 367.8 foot high specimen discovered in 1963 by the National Geographic Society, grows in Redwood National Park (along with the second, third, and sixth tallest trees).

Neither species is the world's longest living tree; that distinction belongs to a patriarch of California's White Mountains—a bristlecone pine—which has reached an age of 4,900 years. Yet the redwoods are not youngsters. A lifespan of 3,000 years is not uncommon for the giant sequoia, and one coast redwood is known to have lived 2,200 years. However, the average age for canopy height coast redwoods ranges between 400 and 800 years.

There are other differences between the two California species. The giant sequoia prefers a Sierra habitat, and will withstand snow and freezing as well as hot, dry temperatures. It grows in more open stands, and its bark is more cinnamon-colored. The coast redwood cannot tolerate temperature extremes, and does best where summer fog and winter rains prevail.

Although coast redwoods grow on both hill slopes and alluvial (streamside) flats, trees along the partially protected alluvial flats tend to grow taller than those which must constantly battle the winter winds that pummel slopes and ridges.

In contrast to their great height and weight, redwoods have a relatively shallow root system. Their roots extend outward perhaps 50 feet, and downward only about 6 to 10 feet. The bole, or trunk, which rises from this root system is usually quite straight, tapering to a spire at the top. Its base is flared, giving the tree additional support.

The life cycle of the coast redwood begins between November and March when tiny male and female conelets are produced on different branches of the same tree. Pollination usually occurs in May. The winds blow the sulphur-colored pollen from the male cone to the female cone. Once fertilized, the female cone matures in approximately five months. The female redwood cones are smaller than a human thumbnail and it takes more than 200 of them to weigh one pound. Their seed is only 1/16th of an inch long, no bigger than a tomato seed, and yet it will produce a tree that may eventually weigh 500 tons and

Last Chance Trail, near the seacoast in Del Norte Coast Redwoods State Park, is tinged with pink rhododendron blossoms in the spring. Here the visitor will also find spruce and alder, fine birdwatching opportunities, and spectacular vistas of the Pacific Ocean.

reach heights exceeding that of the Statue of Liberty.

Once released from their cones, redwood seeds tend to settle within 400 feet of the parent. And though a high wind can sometimes carry them far afield, the seeds usually germinate on duff or logs, in debris, or under vegetation gathered around the parent tree's base.

The first few months of a seedling's life are critical. Like their parents, seedlings require considerable moisture. If rainfall is less than normal, they may die. They must also survive enemies such as slugs, snails, birds and rodents. And during the sapling stage, they may be eaten by bears and porcupines. Only about one out of every million seedlings will become a mature tree.

Coast redwoods also reproduce by sprouting, a rare characteristic of conifers. The sprouts result from the elongation of a dormant bud on the stem. Large woody masses of these form the burl tissue for which the tree is well known. Inside, their wood-grain is intricate and beautiful. Some of these burl growths may weigh as much as 500 pounds, and large or small, they are able to produce sprouts even when removed from the redwood's trunk.

Botanists have determined that sprout redwoods usually grow at a faster rate than seedlings; the latter may take three years to reach a height of 12 inches while vegetative sprouts may grow 15 inches by the end of one year.

Though coast redwoods require considerable moisture, excessively wet, weakened soil is literally their downfall, especially if they are exposed to high winds. Generally, however, these monarchs endure more problems and perils than trees of lesser hardiness. No known diseases kill mature trees. Nor do any insects seriously damage them.

While tolerant of flooding, redwoods cannot endure prolonged periods of saturated soil—conditions which deprive the roots of essential oxygen. Sometimes a tree survives silt buildup by establishing a new root system higher up on the trunk. Yet, at other times, even a deeply-rooted redwood "smothers."

Fire, of course, is a serious danger to seedlings and younger trees, although the large, mature redwood is quite fire tolerant. Examination of the growth rings of a tree which fell in Humboldt County in 1933, showed that it had survived no less than 10 fires in its 1,200 year lifespan.

Unlike many trees, redwoods lack flammable resins which fuel fires in other species. However, fires burning in the understory kill smaller redwoods and trees of other species, especially the thin-barked tanoak and hemlock. These fires remove the duff layer and leave a mineral seedbed where shade-tolerant seedlings, including redwood, will flourish.

Even after its death, the coast redwood remains a vital contributor to its environment. As decay occurs, many micro-organisms and fungi convert its wood into nutrients for other plants. Insects and small animals find food and cover there, too. Seedlings grow on its decomposing wood, and life begins anew.

The redwood, though dominant, shares its forest habitat with a wealth of other plant species, and together they make the forest a place of diverse sights, smells and impressions.

The Douglas-fir is the redwood's closest rival in both age and size. Distinguished by its light brown, deeply-furrowed bark, its needles are very aromatic, and may be used in brewing an herbal tea. Other trees include the acorn-bearing tanoak, western hemlock and madrone, the latter with its distinctive bright red bark which many visitors mistake for manzanita. Near streams, the big leaf maple and alder are found; near the coast, the Sitka spruce is common.

The small plants that form the ground cover in the moist forest include redwood sorrel (or oxalis), wild ginger, redwood violet, bleeding heart, several species of ferns and countless herbs. The redwood sorrel, a most distinctive plant, grows in huge mats which spread across the forest floor. These plants are so well adapted to the shade that their three-part, clover-like leaves fold up when exposed to full sunlight.

Flowering shrubs and berries include dogwood, the California huckleberry, azalea, and the rhododendron. Less appreciated by the visitor, perhaps, is poison oak which often climbs high on redwood trunks. And in spite of its toxicity to humans, it is an important food-source for many animals.

Some visitors notice the seeming scarcity of wildlife in the redwood forest. But if they are sharp-eyed and patient enough, they need not be disappointed. Shrews weave through the forest duff in search of prey. The Douglas squirrel, or chickaree, spirals countless redwood and Douglas-fir trunks, pausing only long enough to feed on their seed-cones. Western gray squirrels haunt the tanoaks, where they eat acorns and fungi. The dusky-footed wood rat, which weighs up to one-half pound, makes huge nests of twigs and leaves; it feeds on seeds, acorns and green vegetation as well as fungi. Bear, elk, deer, coyote, and bobcat also roam the redwoods.

The robin-sized varied thrush, the winter wren, and the osprey are among the few birds observed along the redwood's alluvial flats. Bird life is more plentiful among slope redwoods and in the park's coastal areas where the scolding shrieks of Steller's jays and ravens are sharply contrasted with the musical notes of the thrushes.

Together, the forest's redwoods and other trees, its shrubs, herbs, flowers, and animals form an important ecosystem. Rare in populous California, this varied mixture represents a true *climax community*. This means that a natural balance has been attained by an ecosystem over a long period of time, perhaps even centuries. Left undisturbed by human impact or natural catastrophe, a climax community will perpetuate itself indefinitely, finely-tuned by nature over a timespan that dwarfs man's own brief existence.

What the visitor comes to appreciate and understand in the redwood forest is not just the magnificent coast redwood itself, but the entire forest: complex, delicately-interwoven, balanced. While he is here, man, too, is part of this community.

Sword ferns and clover-like oxalis carpet the base of a moss-covered big leaf maple in Prairie Creek Redwoods State Park (opposite). Rising sharply beside the banks of bubbling Home Creek (above) are the fern-covered walls of Fern Canyon. Many species of ferns, including maidenhair, five-fingered and sword, cling to the walls of this picturesque gorge which was formed by the force of erosive stream action. Visitors exploring the canyon by trail will also see alders, thimbleberries and a variety of mosses.

Typical scenes on the redwood seacoast.

THE SEACOAST

In contrast to the misty, primeval stillness of the redwood forest, the redwood seacoast, at most times, is a place of sound and fury: the crash of wave on the rocky shoreline, the hiss of surf on sand, the wail of a harsh northwesterly winter wind.

But at other times, the seacoast is tranquil. Fog, the same fog which mutes the babble of forest brook, hushes the roar of the surf. In summer and fall, gentler breezes sculpt filigree from the spindrift of wavetops. The very closeness of this vast ocean dwarfs, humbles, and overwhelms the human mind and spirit.

Nothing is static on this lonely, brooding interface of sea and land. Not the shoreline itself, not plant, not animal. Wind-driven waves constantly mold and remold, adding a pinch of sand here, removing a clump of rock there—just as the more predictable tides cover, expose and then re-cover communities of intertidal zone creatures.

The changing, seasonal climate dictates which birds visitors will see; a majority of the park's 300 species are marine, and many of these are migratory. Even the organisms which inhabit the relatively-still river deltas are not immune from change. In winter, when storms send tons of fresh water into river mouths and alter their levels of salinity, animals must adapt or perish.

A massive, semi-permanent weather system called the Pacific High is primarily responsible for the conditions which control this coastal environment. In spring and summer, the high pressure dominates atmospheric circulation, and winds are generally northwesterly. In early fall, the high weakens and is pushed about by stronger lows moving in from the Gulf of Alaska.

Rainfall, fog, ocean winds, wave patterns—all are controlled to an extent by these shifting high and low pressure areas, as is the California Current,

"THE SEA NEVER CHANGES
AND ITS WORKS, FOR ALL THE TALK OF MEN,
ARE WRAPPED IN MYSTERY."
—Joseph Conrad *(Typhoon)*

which moves southward along the coast.

In turn, the current exerts its own influence on the coast. Though it is a cool current with very little seasonal temperature fluctuations, its position shifts slightly due to the movement of the pressure systems. As it moves, it carries rich nutrients up from the ocean depths and provides a food supply for many coastal organisms.

Seashore visitors may view an incredible mix of plants and animals. The myriad habitats range from the offshore region beyond the surf, past the intertidal zone which is sometimes wet, sometimes dry, through the coastal strand and coastal shrub environments, to the coastal spruce forest.

Past the surf, seals and sea lions search for fish and other food. The seals and sea lions are pinnipeds (fin-footed) which feed upon fish and crustaceans. The commonest ones here are the giant Steller's (which may attain a weight of one ton), the playful California sea lion, and the harbor seal.

Telltale spouts mark the presence of California gray whales, migrating between the Bering Sea and Baja California's warm water lagoons. Dolphins and porpoises and killer whales, all cetaceans like their relative, the gray whale, are here too.

Of the coast's seabirds, gulls are the most common species. Cormorants frequent lagoon, river or shore waters. Though the endangered brown pelican is not known to nest in the area, it is a frequent summer visitor. Among other birds are black oyster catchers, willets, black turnstones and sanderlings.

Between the sea and the low expanse of grass, reeds and half-submerged meadowlands lie the sandy bottoms and rocky tidal pools of the park's intertidal zones, a habitat for an extremely productive and varied biological community.

Above and inland are three distinct plant and animal habitats, many of whose populations overlap. Grasses are the most noted feature of the coastal strand and dune and grow on upper beach sands above the summer high tide line. The best examples are at Gold Bluffs Beach.

Where the shore begins to slope upward, coastal shrub vegetation is prolific. All along the coast, this type intermixes with coastal prairie and coastal forest, forming a patchwork of many types of plants. As its name suggests, it is a region of low-growing shrubs—coyote brush, silk tassel, currant, blueblossom, and salal—all woven into a dense mat. Here and there, across the slopes' grassy areas, are the California blackberry vine and clumps of sword fern. Spring and summer brings forth a profusion of color as wildflowers burst into bloom. Here are purple lupine, larkspur, thistle, white seaside daisy and yellow mustard. Many birds and small mammals feed and find protection on these slopes. Commonly seen birds include the the song sparrow, wrentit, Bewick's wren and red-tailed hawk.

The Sitka spruce and red alder forest might be called the dividing line between the coast itself and the redwood forests farther inland. Occasionally, in well-protected areas of this habitat, the visitor may see a redwood or two, but only where it is sheltered behind a guardian Sitka spruce. Far more tolerant of salt air and wind, the Sitka spruce is the most common evergreen growing on the redwood coast.

In windier sections of their habitat, Sitkas and red alder are often shaped and pruned into odd sculptures, graphically demonstrating the power and artistry of wind and spray along this magnificent coastline.

Fog, the bane of the mariner and the curse of the motorist, is nonetheless vital to the life of the coast redwood. On this day, however, it was absent and clear weather prevailed at Endert's Beach.

Warmly aglow in the late afternoon sunlight, wave-polished driftwood lies on a sandy shore south of Gold Bluffs Beach. With the next high tide, perhaps, it will be gone. Opposite: a spectacular expanse of shoreline in Redwood National Park, with Caruthers Cove and Gold Bluffs Beach visible at the upper left.

A redwood grove beside Prairie Creek and (opposite) Espa Lagoon.

RIVERS AND STREAMS

The clear, fresh, bubbling water of a mountain stream trickles down to join a larger creek or river, and together they tumble toward the sea to the west. Where these fresh waters mingle with the salt of the ocean, estuaries are formed. Flowing in from the sea to the land, waves slam the shore, claiming parts of it to form brackish marshes and lagoons where reeds and grasses intermingle, and where wildlife abounds.

These varied bodies of water, large and small, roiled and calm, comprise the riparian system of the redwoods.

Water is the lifeblood of both forest and prairie, and the system that provides their vital moisture—rainfall, evaporation, more rainfall—is an example of nature's orderly recycling and replenishment.

For humans, the park's waters offer tantalizing incentives. One can wade, swim, raft, canoe, or simply partake of some of the West's finest fishing. Trails provide for slow-paced nature studies; streamside campsites grant solitude.

The Klamath is the largest waterway in the region, and though its headwaters lie high in Oregon's mountains, it is a river rich in California history. For centuries, Indians have netted its abundant salmon and from 1879 to 1895, the Native Americans operated a ferry at the river mouth. The Klamath is summer-smooth and winter-raging. Only its mouth lies within park borders, and it can be found about 22 miles north of Orick.

Originating 50 miles southeast of Orick, Redwood Creek flows in a generally northwest direction and enters the park 18 miles southeast of Orick. Supporting some of the park's densest redwood stands, it demonstrates how human impact can upset and alter a once pristine ecosystem.

The Redwood Creek basin is one of the most rapidly eroding watersheds in

Red alders such as these beside Boyes Creek are coastal trees and are seldom found very far inland. Alders are among the first trees to establish themselves in a forest which has been disturbed.

"NOTHING STANDS UP MORE FREE FROM BLAME
IN THIS WORLD THAN A . . . TREE."
—Henry David Thoreau

the world. Logging and related activities have combined with an unprecedented number of severe winter storms—and the outcome has been the sedimentation of Redwood Creek.

The creek's riffled areas, once favorite spawning grounds for salmon and steelhead trout, are now silted in. Pools have filled with sediment and gravel. And as the sediment load increased, the creek grew wider and its waters became shallow.

The scenic Smith River, in Jedediah Smith Redwoods State Park, is the region's second largest river, and eight miles of it runs through the park. Because the slopes here are generally rocky, the river clears quickly following a storm (unlike Redwood Creek where slope erosion is constant), and the Smith's sparkling waters remain prime habitat for cutthroat and steelhead trout as well as king and silver salmon.

Coastal lagoons form another important segment of the riparian system in and near the park. Freshwater Lagoon, located one-and-a-half miles south of Orick, has abundant trout. Just south of the park's boundary are the easily accessible Big and Stone Lagoons; both offer fine fishing opportunities. Because these lagoons contain a mixture of saltwater and freshwater, they support a variety of marine fish, including smelt, silverside perch, herring and flounder.

One need not be a fisherman, however, to savor and appreciate the natural wonders of the park's waterways.

In summer, one may see a spotted sandpiper teetering at the water's edge, watching as a river otter splashes and slides in the mud. Upstream, a beaver diligently builds his winter home by tunneling into a bank. In the evening, the air hums with the sound of insects. A dipper or water ouzel (a bird that walks underwater) searches the stream-bottom for its insect dinner. A great blue heron stands poised and intent—*its* dinner will be that unsuspecting frog or salamander. And that racoon prowling the water's edge is hungry, too, as the crayfish just beneath the surface may soon find out.

As summer fades, and the first frost announces autumn's arrival, rivers and streams begin to mirror the golden yellows and rusty browns of the big leaf and vine maples.

At the Klamath's mouth, silver and king salmon wriggle toward upstream shallows and tributaries to spawn. The female creates a nest in river gravel by swishing her tail, and once the eggs are fertilized by male milt, she covers them lightly with sand. The process is repeated time and again. Their progeny, like themselves, will be born orphans and die childless, for the eggs of these migratory or anadromous species hatch after the parents that spawned them have died.

With the arrival of winter, the mood of the waterworld shifts once again. Streams and rivers begin swelling, rising so high that sometimes the moss-covered willow and maple trees along the banks are bent or ripped loose by the torrents of angry waters.

But spring is a time of rebirth. The receding waters of creeks, streams and rivers begin reflecting the pinks, blues, and whites of azalea, blueblossom, and dogwood. The pungent aroma of skunk cabbage fills the air; and streamside woods, nurtured by winter rains, begin glowing with yellow violet, pink salmonberry, and a whole rainbow of other flowers. With winter's rest at an end, animals begin cautiously poking their heads outside their homes.

Soon, it will be summer again.

Bigleaf maples (above, on Redwood Creek) and alders (opposite) are among the deciduous trees common to areas where moisture is abundant. Many grow on the edge of the National Park's redwood forests.

Symbols of the prairie: open space and wildflowers.

THE PRAIRIE

If the forest could be called Redwood National Park's sea, the prairies would be its islands. Contrasting sharply with the dense, green groves found in much of the park, the prairies are treeless, grassy havens where wildlife abounds; where spring wildflowers are an explosion of color; and where summer mellows to amber.

Little is known of the exact conditions that created and maintained these prairie-meadows. Their soils vary somewhat from those in the forest and soil slippage is undoubtedly a factor. But roadbuilding and the introduction of non-native grasses have also altered the prairies and perhaps even more than natural factors, what the prairies are today is the result of human impact.

In the fall, Native Americans congregated in these airy, sunlit patches to collect acorns in the oak woodlands at the prairie's edge; and in order to make it easier to gather this most important staple, they often burned away the dry leaves beneath the trees.

Prairie margins were also burned in order to stimulate the hazel so prized as basket material. Then, too, fire eventually produced a whole new assortment of elk and deer forage, and with the tender new shoots as bait, Indians soon found hunting more profitable. Deer were brought down with bows and arrows while elk were captured in pits and snares.

Burning also helped enrich the supply of "Indian potato," a plant common throughout the Pacific Northwest. Actually a lily, it was an important food of the Yurok Indians who used sticks to dig the edible bulb from just beneath the prairie's grassy floor.

After the arrival of European settlers, the prairie lands were used for pasture and cultivation. Even today, home-steading remnants are visible—a weathered ranch building here; a

Vivid examples of nature's artistry, these are only four of the park's many flowering plants. Clockwise from the top left, they are the calypso orchid, Indian paintbrush, redwood sorrel (or oxalis) and columbine.

"WE ARE SHAPED AND FASHIONED BY WHAT WE LOVE."
—Goethe

dilapidated water trough or old plow there.

Prairies are scattered throughout Redwood National Park. They are found both near the coast and inland, on slopes as well as in more level areas. Two major prairies lie just beyond the coastal strand of the Gold Bluffs and Crescent Beach areas. The inland prairies are comprised of the Bald Hills, near Orick, and the Little Bald Hills, east of Jedediah Smith Redwoods State Park.

Grasses dominate the prairie, and in locales where there is sufficient moisture and little disturbance, the slim stalks often grow waist-high. Such is the case at Gann's Prairie, near Orick. Northern reed grass is dominant here, but no less than a dozen native species thrive—more than in any other known California coastal grassland area.

Between the prairie and the redwood forest there is usually a woodland rim with its intermingling of oak, madrone, California laurel, big-leaf maple, and Douglas-fir. Near the coast, red alder and Sitka spruce decorate prairie margins.

The Douglas-fir has been a persistent invader of the prairie. Indians once used fire as a means of keeping these aggressive trees in check; the modern absence of fire or other disturbances allowed the firs to take hold along disturbed prairie edges. But because fires have been systematically excluded, the firs have once-again taken hold along disturbed prairie edges. A 1916 account of the prairies near the Klamath River described them as "vast tracts of grassy land." And though unbroken expanses remain today, they would no longer be considered as "vast."

The prairie is one of Redwood National Park's most prolific wildlife habitats, and one of the best places to observe birds and mammals.

Swallows are commonly seen flying over the swaying grasses in search of insects. Chickadees are easily recognized by their habit of dangling from twigs and branches while searching for choice berries and seeds. Hummingbirds, grass-seed-feeding juncos, and several birds of prey find easier pickings here then in the forest's many hiding places. Common too, are the California quail, the western meadowlark, and the turkey vulture, which is sometimes mistaken for an eagle.

Among the prairie animals, the Roosevelt elk is the most imposing, a magnificent symbol of Redwood National Park itself. There are several herds or social groups of these antlered mammals in the park. and even in areas where the groups are not separated by dense redwood groves, they seldom mix.

Once widely distributed in California —from Mount Shasta to the central valley—the Roosevelt elk appears to be a docile creature. But it can be extremely dangerous, especially during the mating season when the male is protective of his harem. The massive animals should not be approached under any circumstances. The best place to view them, at a respectful distance, is at Elk Prairie or Gold Bluffs Beach—both within Prairie Creek Redwoods State Park.

Other prairie mammals include deer, bobcat, bear, raccoon, and porcupine. Coyotes, foxes, and mountain lions are here too. But it takes endless patience to spot them moving about on these grassy islands in the forest.

Animals of both forest and prairie, Roosevelt elk are Redwood National Park's largest land mammals. They are closely related to the Rocky Mountain elk. Perhaps seeking dinner—elk are leaf and shrub browsers—these are in Prairie Creek Redwoods State Park.

The Bald Hills region of Redwood Creek Basin contains the most extensive prairies in Redwood National Park. This pleasant meadow is typical: grass-covered, surrounded by trees, a prolific habitat for birds and mammals in contrast to the less plentiful populations of the redwood forest.

Time passes, but the forest lives on.

HISTORY

If a thousand-year-old coast redwood could talk, imagine the stories it would tell. From its seedling beginnings to its patriarch years, generations of humans have passed by—Indians, trappers, settlers, farmers, gold miners, loggers, dairymen, fishermen and canners, the armed forces of two world wars, and concerned conservationists. Thousands of sunrises have come and gone. And still the redwood lives on, connecting us by a hand-touch to all the generations it has known.

The earliest were the Indians. As their ancestors had before them, the Yurok, Tolowa, and Chilula lived in villages along streams and on the coast. Though the three groups spoke different languages, their cultures were similar.

Most numerous were the Yuroks, who maintained 55 villages along the Klamath River and portions of the coast—from present day Trinidad, north to False Klamath Cove. The Tolowa territory included the Smith River and the California coast from False Klamath Cove northward. The Chilula built most of their settlements in the Redwood Creek Basin.

The redwood region provided timber for the Indians' frame houses, canoes, and implements—along with abundant food supplies from natural sources. Acorns were the main vegetable staple and were supplemented by fish (mostly salmon), which the Indians took either with gill nets or seines. There was a plentiful supply of shellfish as well as deer—and the occasional stranding of a whale provided a serendipitous feast. To add variety, the hunters occasionally harpooned seals and sea lions, which they attracted by wriggling their disguised bodies and imitating the animals' barking.

Sporadic contact with Spanish and English maritime explorers began in the mid-1500s when ships of these expanding empires sailed past the Humboldt coast. A long lull ensued, and it was not until the late 1800s that Spanish, English and Russians came in any number, stopping only occasionally to trade for food and furs. And other than Fort Ross, the Russian's small colony to the south, they built no permanent settlements.

In April, 1828, the famed explorer and mountain man, Jedediah Smith, left California's central valley with a party of 20 men and 300 horses and mules. He was determined to reach the Pacific, and hoped to get there by following the Trinity River downstream. At first, the party made good progress, and by May 19, the ocean was in sight. They pitched camp atop a high ridge covered with fir, spruce, and redwood—in or near the area of present-day Gann's Prairie (near Orick).

The rest of the trip was not as easy. The men found the forest here impenetrable and in trying to find an easier route to the sea, they were forced to retrace their steps to the Klamath River. The party spent another 10 days beating through a maze of undergrowth—all the while bedeviled by both fog and rain. The alternate route had been little easier, but on June 20, 1828, they crossed the Klamath River.

It was 1849 before the next known white man appeared on the scene. Dr. Josiah Gregg, scientist, traveler, and author, led a party of seven men down the Trinity River, headed for the coast. As had Jed Smith before him, Gregg underestimated the difficulty of the trek. His 10-day food supply was not adequate for the four-week journey. Gregg died during the expedition, but surviving members of the party led new adventurers into the Humboldt Bay region, and boom towns began springing up.

One day in April, 1850, J. K. Johnson scooped up a few grains of sand while strolling along the beach north of his home in Trinidad. The sand had a tantalizing appearance. Was this gold? Johnson rushed home to have the sample assayed and to reprovision for a return trip to the beach. To his dismay, he found that the tide had obliterated the location. But it wasn't long before the presence of gold was confirmed at Gold Bluffs Beach.

Copper and silver mining fill other brief pages in the region's history. In the meantime, the industry that was to dominate the entire economy of the redwood region, and in fact to drastically alter the face of that region, was founded. It was logging. The first lumbering operations were recorded in 1850. No redwood was cut commercially in either Humboldt or Del Norte Counties, however, until 1855. The earlier timbermen had not only been confounded by the tree's size, but had lacked equipment large enough to cut, transport or mill it.

For the redwoods, the respite was short-lived. In a few years, thousands of board feet of redwood timber were making their way to ports around the world. By the turn of the century, one logging company alone was harvesting 8.5 million board feet every year, and it owned more than 8,700 acres of forest land. By 1929, the rate of cutting increased to about 500 million board feet annually.

The huge size of the trees triggered a revolution in logging technology, especially in the methods of cutting, transporting and milling. Sometimes it took two choppers an entire week to fell a single redwood. After the tree was felled, teams of bark peelers went to work, and the logs were then dragged to mills by oxen, horses or mules.

As the tempo of logging increased,

"A MAN IS RICH IN PROPORTION TO
THE NUMBER OF THINGS HE CAN AFFORD TO LEAVE ALONE."
—Henry David Thoreau

---

other changes took place. Logging roads were cut, coastal roads were built, and farmers and merchants began to clear and settle the forests.

Little, if any, thought was given to the effect that logging and related activities were having on the environment. The supply of trees seemed limitless. By the middle of the 20th Century, bulldozers, power saws, and modern trucks began replacing hand axes and animal teams. The public became increasingly concerned.

The earliest formal effort to preserve virgin redwood forests dates back to 1852, when California legislator Henry A. Crabb introduced a bill proposing a state redwood park. But it was not until 1902 that such conservation efforts finally bore fruit. That was the year the California State Legislature established Big Basin Redwoods State Park in Santa Cruz County.

In 1908, philanthropist and Congressman William Kent donated Muir Woods, near San Francisco, to the federal government. Designated as a national monument, it was the first involvement of federal government in coast redwood preservation.

The most dynamic, continuing, single moving force behind the eventual establishment of a Redwood National Park, however, was the founding in 1918 of the Save-the-Redwoods League. Under the direction of John C. Merriam, a University of California paleontologist and later president of Carnegie Institute, and Newton B. Drury, who became the fourth director of the National Park Service, the League initiated a drive to raise private funds to buy privately-owned redwood groves for public preservation.

A national depression and World War II intervened, but the work went on, and the League continued to help expand and protect California redwood state parks—grove by grove.

Through the efforts of the Sierra Club and the Save-the-Redwoods League, the idea for a national park gained new momentum in 1961. Aided by a $64,000 grant from the National Geographic Society, the National Park Service conducted a survey to locate the remaining old-growth redwoods and to determine an appropriate site for a park.

The report, released in 1964, stirred widespread public attention. From the original 2 million acres of primeval coast redwood forest only about 300,000 acres remained—and of that, only 50,000 acres were protected within the California redwood state parks.

After considering various proposals, Congress authorized establishment of Redwood National Park, and President Lyndon B. Johnson signed the historic act into law in 1968—the 50th anniversary of the founding of the Save-the-Redwoods League.

The original boundary, however, did not include enough land to insure adequate watershed protection for Redwood Creek.

On March 27, 1978, President Jimmie Carter signed the Redwood National Park Expansion Act. The passage of this legislation indicated that Congress realized the damaging effects that certain land-use practices which occur outside a park could have on park resources. The act added 48,000 acres to the existing park—mainly within the Redwood Creek watershed. It also provided for a massive rehabilitation program of cutover and other erosion-damaged lands. And even though the virgin groves along Redwood Creek would never again be as wide-reaching as in the days before Europeans arrived, at least now some future loss might be avoided.

A far cry from modern logging equipment, hand saws and axes were the main tools when redwood lumbering began in the 19th Century. The redwood's massive base could not be cut efficiently, so loggers usually began cutting six to 10 feet off the ground where the tree was narrower. They positioned themselves on springboards inserted in the trunk (opposite). The steam-operated Dolbeer Donkey (above) first appeared in 1882 on the Salmon River and greatly cut the time required to snake felled trees to haul roads for hitching to oxen. This Dolbeer Spool Donkey, photographed about 1900 in the Fort Dick area, was a much-improved model.

In time, with man's help, nature will heal her scars.

REBUILDING A FOREST

Nature heals her wounds, but it is a long, slow process. And when human impact compounds natural disturbances, the wait can seem interminable.

When Redwood National Park was established by Congress in 1968, more than a century of logging—greatly intensified during the 1950s and 1960s—had substantially reduced the once-vast redwood groves of Humboldt and Del Norte Counties.

Some of the finest remaining stands within the new park's boundaries, including the giants of the Tall Trees Grove, were endangered as a result of continued logging outside and upstream of the park. Most adversely affected was the Redwood Creek Basin where erosion and downstream sedimentation far exceeded tolerable natural levels.

In 1978, Congress passed the Redwood National Park Expansion Act. The legislation added 48,000 acres to the park's original 58,000 acres; at the same time, it established a 30,000 acre Park Protection Zone (PPZ) upstream in the Redwood Creek Watershed. The PPZ is a sort of "buffer" where the park staff works closely with private timberland owners to minimize erosion from timber harvest operations and other land-use activities.

In the same act, Congress earmarked funds for a far-reaching 10-15 year program to rehabilitate the park's cutover forest lands, with a major emphasis on erosion control.

Because of the redwood's immense size and the wet, steep nature of the slopes on which many grow, their removal entails severe ground disturbance. Tractor logging directly alters the landscape and causes soil compaction, loss of topsoil, and destruction of ground cover. Culverts and fills placed in stream channels when roads and skid trails are built cause washouts and landslides. Slopes

"WHAT WOULD THE WORLD BE, ONCE BEREFT
OF WET AND WILDNESS? LET THEM BE LEFT . . ."
—Gerard Manly Hopkins *(Inversnaid)*

undermined during roadbuilding are also subject to failure and landsliding. Where stream channels are filled, further landsliding results as diverted water fills neighboring streams above their capacity.

Many of the Redwood Creek Basin's slopes are just too unstable to accept poorly designed and maintained logging roads. Drainage alterations cause increased downstream and downslope erosion. The naturally unstable nature of the slope increases landslides and causes higher sediment loads and streamflow—which leads to streambank erosion. This jeopardizes plant and animal communities.

About 230 miles of logging roads and 2,500 miles of tractor trails—all of which disrupted certain aspects of the watershed—were recognized as either an existing or potential erosion-source within the newly enlarged park.

From the onset, the rehabilitation program has been geared to speed the recovery of natural forests, stream systems, and life communities. To accomplish that, all unnecessary roads are re-shaped to original contours. Sites unlikely to make a comeback on their own are revegetated. Organic and logging debris are excavated from altered stream channels and the water is returned to its historical course.

Mapping is done during the rainy season, when flowing waters make erosional problems most apparent. In the dry summer months, heavy earth-moving equipment re-contours the land. Bared areas are planted in fall and winter.

From the rough re-shaping of each site to the final revegetation for stabilizing bared areas, many techniques are used to control further erosion and channel the runoff that caused it.

Erosion control devices are installed to dissipate the erosive force of rain and to control sheet, rill, and gully erosion. Waterbars are placed where they will best divert surface runoff onto more stable areas. In most instances, all rehabilitated sites receive heavy applications of straw mulch, the most cost-effective method of controlling surface runoff and minimizing further erosion.

The re-establishment of vegetation is the best defense against erosion on rehabilitated sites. Plant species with fibrous root systems knit the surface soil together. Plants with rhizomes—lateral roots just beneath the surface—bind larger blocks of soil. Plants with large, deeply penetrating roots give the subsoil greater strength. And low ground cover plants reduce the impact of rainfall, as do trees that shed their leaves and provide ground litter.

Despite the fact that many slopes, streamsides and prairies are on their way to recovery, the rehabilitation program is far from over. Between 1978 and 1982, the program progressed from an experimental phase to today's multi-faceted program involving a staff of 25 specialists and scientists.

A century or more may pass before the worst of the logging scars disappear and substantial salmon and steelhead populations return to Redwood Creek. It may be as long as five centuries before the seedlings of today reach the stature of their felled predecessors. But at least a bold step has been taken to help speed the process.

Erosion following a timber harvest created the landscape upset in the top photograph. The damage need not be permanent, however; at the bottom, young Douglas-firs are already growing in an area only recently rehabilitated.

TREES IN TRUST

Time, time as we dissect it in days and hours and minutes loses all meaning in a setting such as this. Here is a forest that was young when life itself was young. Here are trees that have already stood for a millenium or two—and still their lives will outlast yours a thousand years.

—PHILLIP HYDE AND FRANCOIS LEYDET
The Last Redwoods

They are kings—the mightiest of their race—and few visitors strolling through groves of these majestic giants could fail to be impressed or stand in respectful awe. For these trees are survivors. Centuries of wind and flood and fire have come and gone; and still the redwoods stand.

It is not surprising then that in 1980 the United Nations Educational, Scientific, and Cultural Organization (UNESCO) declared Redwood National Park a World Heritage Site. With that designation by the 21-nation World Heritage Committee, the park is bestowed with international recognition; for only properties of "outstanding universal value to mankind" are considered as World Heritage Sites.

Thus, Redwood National Park joins Australia's Great Barrier Reef, Arizona's Grand Canyon, Egypt's Great Pyramid, and more than 100 other natural and man-made wonders as being especially worthy of saving for present and future generations.

Designed and produced by
McQuiston & Daughter, Inc.
Text edited by Sandra L. Keith
Lithographed by Paragon Press, Salt Lake City
Photo credits:
Roger Archibald: front cover
National Park Service: inside covers, pages 26, 35, 38 (Childs Collection), 39, 40, 43
Fred Mang Jr.: title page, pages 16, 34
Save-the-Redwoods League, photo by Howard King: page iv
Richard Frear: pages ii, iii, 3, 29, 33
Woodbridge Williams: page 2, 7
Richard A. Rasp: pages 5, 6, 30, 41
Larry Ulrich: pages 8, 11, 14, 15, 21, 22 23, 24, 27, 34, back cover
Mark Jussel: page 20
Dave Van de Mark: pages 9, 17, 28, 32